After serving as a U.S. Naval Aviator and patrol plane commander in the late 1960s, Richard Wildermann received a master's degree from the Yale School of Forestry and Environmental Studies. He spent his career as an environmental analyst and program manager in the public and private sector, overseeing assessments of major federal projects, including offshore oil development. Mr. Wildermann and his wife, Margaret, live on Seabrook Island, South Carolina, where he is a climate activist.

For Eleanor, Benjamin, Sofia,
and all the world's children.

Richard Wildermann

WILDLIFE ON A WARMING EARTH

Austin Macauley Publishers

LONDON * CAMBRIDGE * NEW YORK * SHARJAH

Copyright © Richard Wildermann 2023

All rights reserved. No part of this publication may be reproduced, distributed, or transmitted in any form or by any means, including photocopying, recording, or other electronic or mechanical methods, without the prior written permission of the publisher, except in the case of brief quotations embodied in critical reviews and certain other non-commercial uses permitted by copyright law. For permission requests, write to the publisher.

Any person who commits any unauthorized act in relation to this publication may be liable to criminal prosecution and civil claims for damages.

Ordering Information
Quantity sales: Special discounts are available on quantity purchases by corporations, associations, and others. For details, contact the publisher at the address below.

Publisher's Cataloging-in-Publication data
Wildermann, Richard
Wildlife on a Warming Earth

ISBN 9781685622053 (Paperback)
ISBN 9781685622060 (ePub e-book)

Library of Congress Control Number: 2023907206

www.austinmacauley.com/us

First Published 2023
Austin Macauley Publishers LLC
40 Wall Street, 33rd Floor, Suite 3302
New York, NY 10005
USA

mail-usa@austinmacauley.com
+1 (646) 5125767

My son and daughter, Nathan and Jodi, and my three grandchildren, Eleanor, Benjamin, and Sofia, were the inspiration for this book. With the steadfast encouragement and patience of my wife, Marg, I can press on with my modest contributions to help resolve the climate crisis. Children everywhere are pointing the way. The world has the solutions to the climate crisis. We just need to rise to the occasion.

Introduction

"Here and there awareness is growing that man, far from being the overlord of all creation, is himself part of nature, subject to the same cosmic forces that control all other life. Man's future welfare and probably even his survival depend upon his learning to live in harmony, rather than in combat, with these forces."

– Rachel Carson (1958)

It's a beautiful day! We had a cold winter. Well, cold for a South Carolina barrier island. Not that long ago the seasons were more predictable. There have always been storms and floods and very hot days in summer and very cold days in winter. But now the storms are stronger, the floods higher, the hot days hotter, and the cold days colder. To give you an idea of what's happening, I'm going to tell you how some of these dramatic weather extremes and natural disasters here and around the world are affecting just some of the creatures in the sea and sky and on land.

My name is Greta. I'm a Carolina chickadee. Now, I know what you're thinking. Carolina chickadees are birds; they can't tell stories. On the contrary, passing on our experiences, our failures as well as our successes, from one generation to the next is essential to our survival. So, story–telling is important. And this story is especially important, not just for chickadees but for all creatures. The natural world that we love and rely on is under grave threat. What's happening is entirely new, something none of the planet's creatures has ever experienced.

Our elders tell us the primary problem is that the air and water are getting warmer. That's causing many creatures to move north to live in places where the temperature is similar to what they experienced in the past. And Jane, my friend the great horned owl, says the warming is also causing the extreme weather events we have to endure more often. I want to tell you all about it because you really must do something.

I must admit, in all modesty, that I'm well qualified to tell this story because birds are very attuned to environmental changes caused by the warming of our planet. Although some birds are very resilient and can adapt to environmental changes, others are more sensitive and respond quickly to even subtle changes in their habitat, including even modest increases in air temperature over long periods. These more sensitive birds can gauge overall habitat quality and will move to places with more favorable conditions. In my short life, I've learned much from even my most distant relatives, and I'd like to pass some of that knowledge along to you.

On this early spring morning, I poked my head out, took a deep breath, and savored the morning sun on my face as it sparkled through the foliage. I noticed Aldo, my gray squirrel pal, sitting on his usual branch in a live oak tree. Squirrels are creatures of habit, but aren't we all? Aldo was excited. He's always excited, his tail twitching. The family with the big house was back and had filled the bird feeders in their yard with all those yummy seeds, including lots of sunflower seeds.

Aldo was clever. People put baffles on the poles to keep him from climbing up to the feeders. Aldo knew he needed birds to come and enjoy munching at the feeders because we dropped many seeds on the ground where he and his furry friends gobble them up. The squirrels have to share the patch of ground under the feeders with Henry, the American crow, and his extended family. But the squirrels and crows get along just fine. The crows are too big to perch on most feeders, so they also dine on the ground. Crows are smart. They always seem to find ways to adapt. And the rest of us appreciate how alert and protective they are. When a hawk is nearby, the crows get excited, make a racket, gang together, and chase the hawk away.

Although it was early spring, I still enjoyed the seeds in the feeders. We visit feeders mostly in winter when insects, our favorite food, are scarce. Aldo was thrilled when I flew to one of the feeders and began tossing seeds left and right. He and Henry shared the bounty of seeds that had collected below.

Living with Nature

"We abuse land because we regard it as a commodity belonging to us. When we see land as a community to which we belong, we may begin to use it with love and respect."

– Aldo Leopold

What is happening to the land and water and sky? So much of nature has been disfigured. Since life began, creatures of every kind have respected their home. That's natural. After all, nature provides us with everything we need: food, shelter, clear streams, and fresh breezes. Why would we abuse it?

The woods on a hillside, a salt marsh tucked behind a barrier island, and a grassy plain stretching to the horizon all have their own heartbeat. The plants that anchor these complex environments and the animals that inhabit them cannot survive in isolation. The vegetation needs air and water and healthy soil to grow and thrive, and then it sustains the resident wildlife, offering shelter and food. The creatures, in turn, realize their dependence on these generous but fragile spaces and treat them with respect.

But humans came along with a different notion. They decided all of nature's bounty was put here just for them, to be dug up, chopped down, and consumed. One would think that after a while people would realize they can't continue to just use up the earth and trees and water and air.

This spinning planet we all share is resilient, but it has its limits. We cannot wear down Earth's natural resources so fast they don't have time to recover. It's not complicated, but humans just don't get it. They're arrogant. They think they can always devise some remedy to put a patch on every injury they cause and everything will be fine. Look at how they cut off mountain tops to tear out rocks that they then burn in power plants that spew poison into the air. Those mountains will never be the same.

While humans crash about, the rest of the planet's creatures tread lightly. We take only what we need and then allow time for the forests, fields, rivers, and oceans to repair the gentle pokes we give them. There is much people could learn from nature if they would just take a moment to look and listen.

"The love for all living creatures is the most noble attribute of man."

–Charles Darwin

Bird Breeding and Food Supply Are out of Sync

"Everyone likes birds. What wild creature is more accessible to our eyes and ears, as close to us and everyone in the world, as universal as a bird?"

– David Attenborough

The weather changed dramatically. We've had almost two weeks of unusually hot days for early spring, with frequent evening thunderstorms. Because of the high temperatures, insects emerged early, providing a plentiful food supply for us chickadees. But we always rely on this bounty when our young are born, and most chickadees are just beginning to lay eggs. It will be more than a week before any nestlings appear, and that is when we need peak food production.

Birds use cues, such as air temperature or hours of daylight, to determine when to breed. When a bird's cues are different than those of its food source, such as insects, a mismatch can occur and the food is not as readily available when the bird needs it most. As a result, nestlings have less to eat and are weak.

In recent years, spring has been coming earlier along with insects and berries. Birds have built-in flexibility to alter their behavior somewhat to adjust to small variations in environmental conditions. So in some places, birds are adapting to this change in the environment by breeding earlier. But birds are having difficulty adjusting to the recent accelerating increase in air temperature, and they cannot adapt to extreme events, such as the excessively hot days we're having now, which are happening more frequently.

The long stretch of hot days so early this spring took its toll. Many nestlings, already weak from a meager food supply, did not survive.

Birds and animals elsewhere in the world live in very different habitats but are suffering similar fates due to warming. The Arctic is a good example. I learned about life in the Arctic from a red knot named Sylvia. Red knots are small shorebirds that migrate thousands of miles twice a year, first in early spring when they leave South America and fly to the Arctic where they have their young. Then they return south in the fall. I met Sylvia recently when I visited our beach, as I do on occasion. I live in the woods, but I've always been curious about birds that live along the shore. Sylvia had selected our beautiful beach to rest for a couple of weeks during her long journey. She and her relatives put their time to good use during their brief visit with us, resting and eating to gain strength before continuing their arduous flight north. The picture Sylvia painted of the Arctic was fascinating.

Because temperature increases are greater in the Arctic than anywhere else in the world, some birds and their food sources are out of sync there as well. Shorebirds that breed in the Arctic rely primarily on insects to feed their young. Insects emerge earlier in years when the snow melts earlier, which is happening more often as the Arctic warms. As a result, chicks of many Arctic shorebirds hatch after the peak period when insects emerge.

The Baird's sandpiper, a shorebird, migrates thousands of miles in spring from South America to the high Arctic and lays a clutch of eggs soon after arriving. Ideally, the eggs hatch a few days before insect production peaks to ensure an optimum food supply when the chicks need it most. But Baird's sandpipers have been breeding earlier in the spring because of higher temperatures, and often their chicks are born before the insects they feed on are most abundant. The result is a reduction in the growth rates of the young birds, and that can lead to population declines.

Arctic seabirds, which reproduce during a very short period in spring when daylight, water temperature, and food availability are just right, are also being affected by changes in their environment caused by warming. In response to increasing ocean temperatures, some arctic seabirds that get their food in shallow waters try to adapt by breeding earlier in the spring.

But reproducing earlier does not necessarily solve the problem because warmer waters can alter the availability of primary food sources for seabirds. Many seabirds are using precious energy flying longer distances to find food for their young, and the fish they bring back are less nutritious. The parents suffer and their babies suffer.

There has been a decline in the population and breeding of the black-legged kittiwake, a common Arctic gull, because there are less zooplankton, an important food web resource, due to rising sea surface temperatures. Zooplankton are tiny organisms that feed on phytoplankton, which are single-celled plants at the base of the food web that drift in the ocean. Small fish feed on the zooplankton and the kittiwakes feed on the small fish.

The black-legged kittiwake spends most of its time at sea, returning to land primarily to breed. Off the coast of Scotland, where kittiwakes had been abundant, warming sea temperatures have reduced the number of zooplankton that is a primary food source for the sand eel, the small fish that kittiwakes in these waters feed to their chicks. Many kittiwake chicks are starving because less zooplankton means fewer sand eels. Environmental conditions and kittiwake breeding success vary year-to-year, but the overall decline in some kittiwake colonies is related to a mismatch between its primary food sources at the critical time when it breeds.

Black-legged kittiwake
photo by Lisa Hupp, USFWS

For many birds around the world, it will become more and more difficult to find food for themselves and their young because the warming of the air and water is disrupting when and where their preferred diet is available. Some birds will seek other food while others will alter the time or place they breed and reside. But in the coming decades, some bird populations will decline, perhaps dramatically, because they cannot adjust to the changes to their life's rhythm caused by rapidly rising temperatures.

King Tides on the Beach

"The restlessness of shorebirds, their kinship with the distance and swift seasons, the wistful signal of their voices down the long coastlines of the world make them, for me, the most affecting of wild creatures."

– Peter Matthiessen

One day in late spring we experienced another of those extreme high tides. I always can tell when this happens because water backs up onto the impenetrable roads people create to get around on. These floods are happening more and more often. I thought about Rachel, the least tern I'd met last year during another of my visits to the beach. Rachel had mentioned then that extreme high tides often flooded least tern nests.

Least terns spend the winter more than a thousand miles to the south, and they return north in spring, some to our barrier island. Unlike Sylvia, who visits us briefly to rest during her grueling migration, Rachel stays through the summer. Least terns build their nests, if you can call them that, on the beach just above the ever-shifting strip where the tide comes in and out twice a day. The terns simply scrape a small hollow in the sand, and that's their nest. It's pretty basic and vulnerable.

I wondered if Rachel had arrived and if the day's extreme tide had affected her. So as the tide was receding, I flew the short distance to the beach and found Rachel and many other least terns quite distraught. Apparently, this tide was especially high because of a combination of factors, including strong onshore winds. Many of the adult females had laid eggs a day or two before, and they had been completely inundated. All the flooded eggs were lost.

Fortunately, Rachel had learned from floods in previous years and had scraped out her nest well above the high tide line. Least terns are adaptable. They have learned to find less exposed locations for their nests. Some even build nests on the flat roofs of big buildings built by humans. Those nests have their own problems, however.

In addition to flooding, least terns have to deal with other dangers. Predators, including raccoons, vultures, and great horned owls, attack their nests if the parents leave them exposed for even a moment. The parents are very protective and screech and dive at their attackers.

Intense heat can also take a toll on least tern nests and young birds. The parents will dip their bellies into the water along the shore and then sit on their eggs or young to cool them off.

As the sea level rises, king tides will be higher and occur more often. The least tern nests will be more susceptible to inundation, and the fragile beach habitat used by numerous birds at various times throughout the year will be flooded more frequently. Some birds will adapt by relocating, but in the long term, entire populations could disappear.

Least tern with chicks
photo by Kaiti Titherington, USFWS

King Tides in Salt Marshes

"The marsh, to him who enters it in a receptive mood, holds, besides mosquitoes and stagnation, melody, the mystery of unknown waters, and the sweetness of nature undisturbed by man."

– William Beebe

The extreme high tides that overrun beaches also flood salt marshes, which are an essential ribbon of life between the land and sea.

Coastal wetlands are important habitat for birds, fish, and other animals. Most saltwater fish begin life in salt marshes. Adult fish lay eggs in the marsh, which then serves as a nursery for the young fish. The tide brings in nutrients that nourish salt-tolerant marsh grasses. The grasses grow, then die off and decompose, providing organic material for tiny marine animals that become a food source for small fish. Further along the food chain, larger fish and birds eat the small fish as well as the tiny organisms.

Salt marshes are a major component of the large barrier islands along the southeast coast as well as the smaller back-barrier islands behind them. Because these islands and their associated marshes are very low in elevation, they are susceptible to flooding from even minor increases in sea level and tidal surges. Therefore, many miles of wetlands are flooded from even slight increases in sea level.

A salt marsh may look healthy while under stress from rising seas and more frequent flooding. But then it passes a critical threshold and dies off rapidly, creating open water over vast distances. If that were not enough, salt marshes that are already in a weakened state are more susceptible to other stress factors. For example, a hurricane could devastate huge areas of a debilitated marsh that, if healthy, would be resilient and better able to recover. To add to the problem, the warming air and water the planet now suffers from are creating more powerful hurricanes that inflict ever-greater damage to coastal wetlands.

The loss of marshes can affect adjacent land. As the water level rises, the marshes creep inland in many locations and in doing so they displace the vegetation and wildlife living in that unique transitional zone at the water's edge. Because salt marshes tolerate daily fluctuations in water level as the tides ebb and flow, they are able to adapt to slow increases in sea level by this inland migration and by trapping sediment that gradually raises the marsh mud where the grasses are rooted. However, two factors are working against the marshes. The sea level is now rising too fast in many places for sediments to build an adequate bed for new vegetation, and extreme high tides and storm surges are becoming more frequent. This increase in flooding is causing marshes to die.

When the marsh grass dies, all the animals that depend on it suffer as well. As fragile barrier islands and their marshes shrink due to rising water and erosion from flooding, there is less habitat available for the animals that rely on them. In time, it becomes a matter of overcrowding and the number of animals will decline. There will be fewer small animals in the marsh mud to feed the small fish, birds, and other animals that live in or visit the marsh. Shrimp, crabs, and larger fish, such as flounder and red drum that spawn and grow in the marsh, will also diminish in numbers.

When I explained this to Aldo, he was not impressed. The salt marsh hugging the shoreward side of our barrier island may be important to other creatures, but he never spent time there, so he wasn't concerned. I pointed out to Aldo that everything on our planet is connected to everything else in some way. You don't have to live in a salt marsh, or a rain forest, or the frozen north to appreciate that all these places matter to everyone.

There's a reason warming from greenhouse gas emissions is called global. It's not just happening on our island but around the planet. Earlier I mentioned birds in the Arctic. Let me tell you a little more about the effects of warming in the Arctic.

Melting Arctic Sea Ice

"The Arctic is among the least understood places on the planet; however, we do know that its landscape is changing and evolving as quickly as cell phones and the Internet."

– Phillippe Cousteau, Jr.

When I spoke with Sylvia several months ago, she explained that while Arctic birds are trying to cope with changes in food availability, which is especially critical during breeding in spring, wildlife in the Arctic faces other challenges brought on by global warming. Sea ice is melting at an alarming rate, which affects not just birds but fish and other animals as well.

Sea ice forms and floats in ocean waters of the Arctic and Antarctic regions. Because these two polar regions have significant geographical differences, my discussion will focus on the behavior and importance of sea ice in the Arctic.

The percentage of Arctic waters covered by sea ice fluctuates with the seasons and varies from year to year. Sea ice forms during the fall and winter and reaches its greatest extent in March. Then as temperatures rise in spring, some of the ice melts until September when the cycle starts again.

Because sea ice has a bright surface, it reflects most sunlight back into space. More ice means less warming of the ocean's surface. When some of the ice melts in summer, more dark seawater is exposed to sunlight. Consequently, the water absorbs even more of the sun's energy and melts more ice. This cumulative warming and melting process affects the Arctic environment and even influences the global climate. The accelerating ice melt that is now occurring is a primary reason the Arctic is heating faster than other regions. In just the last few decades the thickness and extent of Arctic sea ice have declined significantly.

This rapid melting of sea ice poses serious problems for the Arctic environment. Sea ice plays an important role in the lives of many mammals and birds that inhabit the Arctic all or part of the year. The rapid decline in sea ice in recent years is altering the Arctic food web and harming to varying degrees many creatures along the coast as well as in the sea.

Algae that grows on the bottom of sea ice forms part of the base of the marine food web in the Arctic. As the ice melts during the summer, the algae is released into the water, attracting zooplankton, then fish, and eventually whales, seals, and birds. Less ice means fewer algae, which affects all these animals up the food chain.

Walruses typically forage for clams on the seafloor, using sea ice in shallow water near shore as a platform. The increase in melting in summer has caused the ice to recede further from shore into deeper water where clams are not available. In response, walruses are congregating on beaches, which is not their preferred habitat.

Polar bears live extensively on sea ice. With less ice available the bears are losing critical habitat, and they have to swim greater distances for food. As a result, they are losing weight and producing fewer cubs, and the cubs are often underweight.

Polar bear with cub
photo by Scott Schliebe; USFWS

Several seal species, including the harp seal, spend most of their lives on sea ice. Harp seals are by nature dependent on sea ice for their survival. In late winter and early spring, females give birth and care for their pups on the ice. They rest on the ice when not in the water feeding on fish. Like some other seals, harp seals have evolved to adapt to modest seasonal fluctuations in sea ice extent, but in the long term, as sea ice continues to decline year-round, it's questionable whether they will be able to make a huge adjustment such as moving to land.

As the Arctic environment is transformed in response to the accelerating rate of melting sea ice, birds and marine mammals are scrambling to find different food sources, increasingly scarce places to forage and rest, and new places to produce offspring and raise their young. The unprecedented loss of sea ice in just the last few decades has global implications. It is contributing to changes in atmospheric circulation and ocean currents and is likely accelerating global warming.

Melting Permafrost in The Arctic

"We need to save the Arctic, not because of the polar bears, and not because it is the most beautiful place in the world, but because our very survival depends upon it."

– Lewis Gordon Pugh

The damage to the Arctic environment from melting sea ice is serious enough, but thawing permafrost makes matters worse. Permafrost is ground that has been frozen for at least two years; in some places on the Arctic tundra, it's been frozen for tens of thousands of years. It can be a few feet or thousands of feet thick. Almost a quarter of the northern hemisphere, about nine million square miles, is covered with permafrost.

When plants and animals in the Arctic die, they often freeze before decomposing completely. As a result, carbon dioxide and methane, two primary greenhouse gases that are byproducts of decomposition, are not released into the atmosphere. This frozen organic matter can build up, trapping the greenhouse gases underground indefinitely as long as the permafrost remains frozen.

But now global warming is thawing permafrost around the world. In the Arctic, where temperatures are rising faster than elsewhere, the rate of permafrost thawing is increasing. When that happens, the organic matter that has been frozen for years decomposes and releases carbon dioxide and methane. Although methane does not remain in the atmosphere as long as carbon dioxide, it traps much more heat. These released greenhouse gases further heat the atmosphere, creating a feedback loop, which further accelerates the warming of the air, land, and sea.

Permafrost is a primary building block of the Arctic tundra. It determines the habitat for a wide variety of wildlife. Vast expanses of the Arctic underlain by permafrost are covered with berries, lichens, and other plants that are essential food for numerous species. Rapidly thawing permafrost is dramatically modifying the Arctic tundra and altering the habitat, behavior, distribution, and abundance of countless animals that live or breed there.

Melting of the ice that holds the soil and organic matter together causes erosion and slumping of the land surface. Lakes quickly drain, shorelines collapse, and silt and pollution, such as mercury, are flushed into surface waters.

Polar bears, caribou, and countless other animals that depend on the more stable land covering the permafrost are deprived of food and water in places where the solid ground literally crumbles away. Changes in tundra snow cover and firmness also affect the ability of mammals to travel and access suitable habitats. Less solid ice and more open water can restrict the movement of larger mammals such as caribou, muskox, wolves, and foxes, which alters their foraging behavior and distribution.

Small mammals such as lemmings, voles, and shrews remain active all year in order to retain sufficient heat to survive. They rely on the insulating properties of snow for shelter in winter and eat plants under the snowpack that also offers limited protection from some predators.

Avian predators such as rough-legged hawks and peregrine falcons build nests on steep slopes and cliffs that are particularly vulnerable to slumping and collapse. Thawing also drains wetlands, the breeding sites for many shorebirds and waterfowl, and ponds where ducks forage and rest.

Melting sea ice and thawing permafrost are dramatically reshaping all segments of the Arctic environment, and that is having ripple effects around the globe.

Drought in California

"Forests are the world's air-conditioning system – the lungs of the planet – and we are on the verge of switching it off."

– Charles, Prince of Wales

One humid summer day I had a chat with Jane, my great horned owl friend. Jane may look fierce, perched motionless high in a live oak tree, but when she's not searching for her next meal, she's a sweetheart.

One of Jane's cousins on the Pacific coast told her that while melting sea ice and thawing permafrost are causing havoc in the Arctic, in California and elsewhere, increasing temperatures are creating a very different problem: drought.

A drought is an extended period of dry weather with little precipitation that causes a shortage of water on or below the earth's surface and damages an area's ecosystems. Although droughts have occurred throughout history, global warming is contributing to more frequent and longer-lasting droughts by altering circulation patterns in the atmosphere, which results in less rainfall in some regions. Warmer air also increases evaporation and transpiration, further drying soil and vegetation, shrinking lakes and ponds, and reducing groundwater recharge.

Droughts harm wildlife by fundamentally damaging habitats and forcing animals to travel further in search of food and water. If a drought is not too severe, some animals can find temporary refuge in small areas that retained suitable habitats. Small animals are less able to move long distances and suffer most. If conditions remain exceptionally dry for an extended period of time, vegetation that many animals rely on to feed their young will not be available and populations will decline.

Trees become weakened from lack of water during droughts. Stressed pines are more susceptible to infestation by insects such as pine beetles, and the infected trees ultimately die. When millions of trees died in California forests several years ago during a drought, it deprived many small animals of shelter and food they needed to survive. These same animals were also more exposed to predators because the drought killed foliage and diminished tree canopies as well as ground cover.

Large mammals can leave a drought-stricken area to find food and water elsewhere. But they will have to compete with other animals and probably people in their new locations. Their populations may dwindle if there is insufficient suitable habitat following a severe drought.

Fish aren't off the hook during a drought. Mountain snow, or snowpack, is the primary source of water for rivers and streams in the west. Fish rely on the melting of the snowpack in spring and summer to maintain stream flows with cool water. While snowpack varies from year to year, over the long-run global warming is reducing snowpack in western mountains. Conditions are worse during a prolonged drought because the snowpack is not adequately replenished during the winter, so there is less cool water during the spring and summer. This leads to low stream flows and high water temperatures which stress fish populations.

More severe, longer-lasting, and damaging droughts are not unique to America. Droughts have long been one of many forces that shape ecosystems in Africa, but in recent years they have been so extensive they are testing the limits of sustainability of some environments.

Drought in Africa

"We are called to assist the Earth to heal her wounds and in the process heal our own – indeed, to embrace the whole creation in all its diversity, beauty, and wonder. This will happen if we see the need to revive our sense of belonging to a larger family of life, with which we have shared our evolutionary process."

– Wangari Maathai

The timing of the rainy season in Africa varies from region to region. In a good year, the wet season provides enough rainfall to sustain vegetation, replenish surface water sources, and ensure adequate soil moisture during the following dry season. However, cyclic changes in ocean currents and winds in the eastern Pacific bring about an increase in surface water temperatures that have historically contributed to periodic droughts in southern Africa, where the rainy season generally ends in March. In recent decades rising atmospheric temperatures due to greenhouse gas emissions appear to be a factor in exacerbating changes in weather patterns, and this has significantly reduced precipitation in southern Africa, causing more frequent, prolonged droughts, with some lasting, not months but years. The severe droughts resulting from these changes in the timing, duration, and magnitude of the rainy season are generating significant reductions in the distribution, abundance, and diversity of wildlife, and lead to increase crowding and disease.

African elephants

Large plant-eating animals, such as elephant, buffalo, giraffe, large antelope, zebra, wildebeest, and hippopotamus, are particularly at risk because droughts reduce the available vegetation and aquatic habitat for these water-dependent species. Mortality from starvation and lack of water is not an isolated event for these animals. As many as 1500 elephants and many other animals died during a drought in the early 1990s, and over 200 elephants died in 2019 in a national park in Zimbabwe. There was a similar incidence of large numbers of elephants dying in the savanna in southern Kenya during a drought in late 2009. As in every ecosystem, an injury to the environment in a savanna ultimately touches all the inter-related biological members in the affected area. Without adequate vegetation and water, herbivores die, and that leads to a reduction in the number of predators.

It is important to remember that droughts are a normal occurrence in Africa. They are a central factor in the natural functioning of complex ecosystems. Greenhouse gas emissions do not cause droughts but there is every indication that the increasing atmospheric temperatures caused by these emissions are bringing about more severe, prolonged droughts that could at some point upset the balance that has prevailed and push some animals to extinction.

Wildfires in California

"I want you to act as if our house is on fire, because it is."

– Greta Thunberg (2019)

While droughts severely stress the environment and wildlife, they also are a contributing factor to larger and more intense wildfires. I needed to learn more about wildfires, so I consulted with Jane, that wise old owl.

Jane has relatives just about everywhere in North America. She knows as much about wildfires as droughts. In fact, last year one of her distant cousins was found debilitated and trudging through the ashes of a California wildfire. He was rescued by a couple of brave – and I must say quite handsome – firefighters. Let me tell you just a little of what Jane's west coast kin have to say about wildfires based on their firsthand experience.

Wildfires are a natural occurrence in forests and woodlands. Over time, wildlife in many places has come to rely on wildfires to regenerate its habitat. But people have disrupted the normal order of things. For years in California and elsewhere, they have created the conditions for many of the wildfires that have been so destructive. In some cases, people started the fires, intentionally or not. And now greenhouse gas emissions, another human phenomenon, are making matters worse. In the past few decades, the number and intensity of wildfires in California and the size of the areas burned have increased dramatically. Scientific evidence shows that one of the main causes is the warming of the air and land surface due to emissions of carbon dioxide and other greenhouse gases.

Wildfires are most prevalent in summer and fall. In the American west, higher spring and summer temperatures are causing the snowpack to melt sooner, which causes the soil and vegetation to dry earlier and remain dry throughout the summer. The drier forests and woodlands are stressed and suffer more readily from infestation by bark beetles and other insects. Large swaths of forest die as a result, making the trees highly susceptible to fire. In southern California, offshore winds increase in the fall causing additional intensification and spreading of fires.

Wildfires affect animals by either disrupting their habitat or harming them directly. Different kinds of animals react differently to a wildfire. More mobile animals, such as large mammals and, of course, birds, can flee. Others, such as ground squirrels and chipmunks, may survive less intense fires by burrowing underground.

Animals and birds that escape a fire often return once the area provides the food, water, and shelter they require. Wildlife will return more quickly after a forest fire that primarily burns undergrowth with minimal damage to mature trees than after an intense fire that destroys virtually all vegetation. However, it is difficult for small animals, such as mice, rabbits, and squirrels, to re-establish without recovery of the ground cover lost in the fire, and larger predators will stay away until their prey returns.

As one would expect, the more intense fires are the most damaging to habitats and, therefore, the most disruptive to animal communities. In one massive, fast-moving fire in California several years ago, thousands of cottontail rabbits perished because they could not move out of the fire's path quickly enough. Some animals that escape may never return to the site of a severe fire. Some will crowd into similar habitats adjacent to the burn site while others, such as mountain lions, can adapt to new environmental conditions.

Mountain lions require large areas. More numerous fires may leave discrete unburned areas that are suitable habitats for mountain lions but are too small to sustain them. Because they are large and very mobile, mountain lions are often able to escape wildfires, but some have been found dead in burned areas following fires in California. Mountain lions are adaptable, and if displaced by a fire they can move into a new habitat and find new prey. However, if an area experiences repeat fires, it may convert to annual grasslands, which is not suitable for mountain lions.

Most adult birds fly away in advance of a fire. While wildfires can occur any time, they are most common well after breeding in spring, so nests and young usually are not affected. But habitats can take years to recover, and returning birds may not find adequate nesting sites the spring following a fire.

The characteristics of a fire are a major factor in determining if and when birds may return. The larger, more frequent, and more intense fires that are occurring in California due to global warming could be having more negative, lasting effects on birds. These fires are more likely to bring about long-term alterations in habitat types, which would change the distribution of bird species. And it's not just resident bird populations that are affected by fires. Migrating birds can suffer if resting spots they rely on along their route are burned.

Wildfires in Australia

"Our spirituality is a oneness and an interconnectedness with all that lives and breathes, even with all that does not live or breathe."

– Mudrooroo

California isn't alone in suffering from wildfires made more extreme by global warming. There have been fires in the Arctic and even in the Amazon rainforest. Many of the rainforest fires were started by people clearing land, but excess warming of the atmosphere has created drier conditions and literally fanned the flames, intensifying the fires. But the fires I want to talk about happened during the 2019–2020 fire season in Australia.

Wildfires had been raging in southeastern Australia for months when much-needed rain finally arrived in January 2020, bringing some relief. By early February, as the rains continued, many fires were being contained.

The devastation in the wake of the fires was staggering. As much as 40,000 square miles burned. There are estimates of hundreds of millions of dead animals.

The conditions that set the stage for these wildfires are all too familiar. The last 20 years in southeastern Australia have been the hottest and driest on record. Shifts in ocean currents periodically reduce rainfall in Australia, but the warming atmosphere is almost certainly a contributing factor to recent longer droughts.

A wide variety of eucalyptus trees make up the majority of Australia's forests. While these trees rely on fires to propagate, the extreme wildfires destroyed many of the trees. Because of the intensity of these fires, the earth and vegetation weren't charred, they were obliterated. This destroyed elements of wildlife habitat critical to some species, such as hollows in trees and deep layers of decaying leaves, that could take decades to recover.

Estimates vary but hundreds of millions of mammals, reptiles, and birds may have either starved, burned to death, or been eaten by predators. Koalas spend most of the day in eucalyptus trees, and the leaves are their primary food. They will instinctively move higher in the trees when threatened by a bushfire, making them vulnerable to intense heat and flames. Koalas are sedentary and slow-moving, and there are estimates that thousands could have perished in the fires. Fortunately, many of these iconic marsupials lived outside the burned areas and survived.

Koala
Pixabay

Fast-moving animals such as kangaroos could have initially escaped the fires, but many died later because they could not find food, water, or shelter in the burned landscape. Snakes and burrowing animals, such as wombats, may have found shelter as the fire approaches by going underground, but they emerged to an uncertain future.

Thousands of dead birds washed up on beaches after they flew out to sea to escape the fire and died of exhaustion or smoke inhalation. Included in these tragic deaths are some of Australia's familiar, colorful birds such as yellow-tailed black cockatoos, rainbow lorikeets, and crimson rosellas.

rainbow lorikeet
Pixabay

The fires posed the risk of extinction for some endangered species whose distribution was confined mostly to the burned areas. These are species you probably never heard of, such as the long-footed potoroo, a small marsupial, and the eastern bristlebird, both of which are unique to southeastern Australia.

Over tens of thousands of years, many of Australia's plants and animals have adapted to periodic fires. But with global warming, the fires are more frequent, more intense, and are decimating larger and larger areas. These fires made it almost impossible for many animals to adapt and have severely tested the resilience of ecosystems.

Some of the fires also overran areas that have not burned for centuries, such as the edges of rainforest in Queensland. Many of the animals in this environment, including unique birds such as Albert's lyrebird, the logrunner, and the rufous scrub-bird have no tolerance for fire because their habitat has been wet for millions of years. Recovery will be slow, or it may not occur at all. What was a wet forest ecosystem with a rich wildlife may be permanently reduced by the fires to dry scrub.

Most people have not taken global warming seriously for various reasons, one being that the damaging effects have not been that obvious. An inch of sea level rise, more frequent king tides, or melting sea ice are not overtly alarming. But the catastrophic wildfires in southeastern Australia are another matter, and they may be a portent of what lies ahead for many regions of the globe. Videos and descriptions of the fires horrified people worldwide. In January 2020 the government pledged millions of dollars for wildlife and habitat recovery. Yet so far nothing meaningful has been done to slow global warming. Recovery is important, but unless the cause of the problem is remedied, the same catastrophes will be repeated. If those fires can't bring about decisive action, what will?

Hurricanes

"As you warm the climate, you basically raise the speed limit on hurricanes."

– Kerry Emanuel

Back on my barrier island, the day began like many others in late autumn. The air was still, warm and humid. High wispy clouds were streaming in from the southeast. Yet something was off. I felt those slight aches that come on just before large storms.

When I looked out from the dense shrub where I'd spent the night, Henry was just returning to his roost. He had been to the beach, and I asked if he had noticed anything foreboding. He said there was only a light breeze but large swells were rolling onto the beach. Some birds are quite sensitive to rapid changes in air pressure, and the seagulls were flying low and not venturing far offshore, which is typical behavior before a big storm. Many shorebirds were moving inland in anticipation, taking cover in dense vegetation. We decided to take extra precautions.

As the day progressed the wind picked up, blustery and erratic. Bands of dark clouds came in waves, bringing bursts of heavy rain. Many birds were leaving, seeking protection further west. Many chickadees planned to hunker down in tree cavities or dense growth, hoping to wait out the coming storm. But first, like most of our friends, we needed to eat more than usual today because there would be less food available immediately after the storm passed.

At night the wind and rain intensified rapidly. The wind became a constant roar. The trees groaned and shook as if terrified. The rain blasted through the vegetation where our flock huddled.

Early in the morning, the rain was easing. The wind had shifted from southeast to northeast and was not as fierce as during the middle of the night.

By mid-day, it was safe to venture out. The damage was extensive, but we were lucky. The hurricane had not made landfall and had rushed north just off the coast.

But there was still considerable damage. Debris was everywhere. Many trees were stripped of their leaves and lost limbs. Some large trees, especially water oaks, had been torn out by the roots or snapped off at their weakest point. There was an eerie silence, no bird chatter, no signs of life.

I flew over to Aldo's nest. There was no sign of him and little was left of his home. Squirrels often suffer major losses, especially the young, during hurricanes. I hoped Aldo would show up safe soon.

I came upon Henry strutting along on the littered ground, his tail swinging back and forth, looking for something, anything to eat. He had waited out the storm in the interior of a shrub, low to the ground, in one of the slightly higher spots on our island. So he had avoided the minor flooding in the island's interior.

The birds and other animals that stayed and weathered the storm began to emerge during the day. Some birds had taken shelter in tree cavities and were lost when the tree was blown down or snapped during the storm. The days ahead would be difficult. Much of the habitat for the island's resident wildlife had been severely bruised. Food would be scarce. Populations of many animals would decline because for years to come there would be less suitable space for them, less habitat offering the food and shelter they required.

Several days after the storm I saw Aldo, exhausted, hungry, but alive. Many of his friends had not survived. At the beach, some shorebirds that had been swept away by the hurricane began to struggle back. Others never returned. The dunes and grasses had been battered by the waves. The storm surge, however, had come ashore hours before high tide, so the habitat of so many birds and other animals along the shore had been spared catastrophic damage. It could have been much worse.

As they evolved, birds and other wild animals acquired survival skills, learning to adapt to disruptions to their habitat. Those skills are being tested as hurricanes are becoming more powerful, with stronger winds and heavier rain. Jane, our great horned owl friend who seems to know such things, says the warming air is causing these more intense storms. Warmer air holds more water, resulting in an increase in rainfall in recent hurricanes. At the same time, the warming ocean water provides the hurricane with more energy, producing more severe and damaging winds.

Most of the severe weather events and extreme conditions such as droughts, floods, and hurricanes that are becoming more routine can all be traced back to one cause, the rapid increase in global warming.

Saltwater Fish

"Even if you never have the chance to see or touch the ocean, the ocean touches you with every breath you take, every drop of water you drink, every bite you consume. Everyone everywhere is inextricably connected to and utterly dependent upon the existence of the sea."

– Sylvia Earle

One mild winter day we had another one of those extreme high tides. It reminded me of Rachel and how so many of her least tern friends lost their eggs when a high tide decimated their nests last spring. I hadn't seen Rachel since she flew south for the winter several months ago. I decided to take a quick swing by the beach to see if today's high tide was washing over the small dunes above the high water mark, and sure enough, once again the water covered the dune grasses. Fortunately, there were no bird eggs or loggerhead sea turtle nests to damage at this time of year.

Looking out at the large expanse of ocean, I recalled Rachel saying that many kinds of fish were being harmed by changes in seawater due to the warming of the air. That concerned her because small forage fish are the most important part of a least tern's diet as well as a primary food source for other birds and larger fish. I worried that Rachel's primary diet may be in danger.

Rachel had explained to me that what's happening in the oceans due to the warming of the atmosphere is complex. I'll try to recall for you some of the problems she described.

Much of life in the sea, including fish, cannot tolerate a prolonged, relatively rapid increase in water temperature. Temperature controls the rate of chemical reactions in a fish that produce the energy and molecules needed to sustain life. Warmer water increases the rate of these reactions, which can affect important fish activities such as digestion and swimming. Warmer waters also cause stress to eggs, larvae, and juvenile fish.

Fish also suffer because the warmer water does not hold as much oxygen, which is as essential for fish and other marine life as it is for land animals. Oxygen in the air dissolves in seawater at the surface. Adult fish can suffocate and fish eggs can die if there is not enough oxygen in the water.

Oxygen is distributed throughout the water column when surface water mixes with deeper water. However, as the surface water warms it becomes more buoyant, and the warming air is disrupting ocean currents. These conditions compound the problem by reducing mixing within the water column, which in turn reduces the amount of oxygen reaching marine life well below the surface. There are now hundreds of ocean areas around the globe where oxygen has been depleted to such an extent that the balance of marine life is severely disrupted.

Many fish can adapt to small, gradual changes in water temperature, but they can't adapt in many areas where the water temperature is rising too quickly and oxygen is depleted. Even if a particular fish can adapt to some increase in water temperature, if its primary food is smaller fish that cannot adapt, then the larger fish is in trouble.

Because of these two conditions – warmer water and less oxygen – many fish will die if they don't move to areas where the water is not warming so quickly. The populations of some types of larger fish are shrinking while others above the equator have moved north into cooler waters. Warming of the air and sea is creating more zones in the oceans with significantly reduced oxygen levels, and these areas are expanding in size around the globe. This is beginning to have devastating ecological effects on all the world's oceans.

All saltwater fish are connected to other living organisms in the sea. The complicated relationships between fish, their living space, and their food sources are intertwined. So these changes in water temperature and oxygen levels are not just affecting a few fish. They affect the entire marine environment.

Heat Waves and Hot Spots
In the Ocean

"In recent years, scientists have documented countless species shifting their ranges toward the poles, higher into the mountains, and deeper into the seas in response to the changing climate."

– Sonia Shah

The oceans aren't warming uniformly or at a steady rate all over the planet. Because of changes in ocean currents and wind patterns, some patches of the ocean are warming faster than elsewhere. Heatwaves are occurring in various places in the world's oceans just as on land. Unusually warm ocean water affects the entire marine ecosystem, including causing algal blooms that are toxic to marine life.

Hot spots form in the ocean and usually last a short time and do not cover large areas. But a few years ago an extreme heatwave occurred in the eastern Pacific. It started off the southwest coast of Alaska in winter and then spread south along the entire Pacific coast to Mexico. It lasted for several years.

A combination of factors apparently caused this extended period of unusually warm water to cover a large area. Warming of surface waters combined with changes in winds reduced upwelling that typically brings nutrient-rich cooler water up to the surface. The result was a profound disruption of the marine food chain. Zooplankton had less to eat for a long time, so animals that rely on them for food suffered. Seabirds and sea lions starved; dead fin whales and sea otters washed up on beaches; Chinook salmon and other fish species were devastated. It will take years for these diverse creatures to recover.

Sea Lions
photo by Lt. Elizabeth Crapo, NOAA

Changes in water temperature and currents are affecting marine life in the northeast as well. Warmer air is hastening the melting of the Greenland ice sheet, and that has slowed the currents that normally bring cooler water to the New England coast. This is causing the temperature of the Gulf of Maine and adjacent waters to increase more than many other places around the globe. The effect on the marine food chain is similar to what happened off the west coast. Plankton is less productive in warmer water. Because of lower oxygen levels, reduced food supply, and warmer water, the populations of some fish are shrinking while others have moved north into cooler water.

The Tasman Sea off the southeastern coast of Australia is another ocean hot spot. In the last few decades, it has heated up several times more than the average increase in global ocean temperatures. Two heat waves occurred in recent years, covering large ocean areas surrounding Tasmania. The first lasted eight months, much longer than previously. As a result, most of the giant kelp that provides habitat for much marine life along the east coast of Tasmania died.

Common kelp, more tolerant to warmer water, began to replace the dying giant kelp. But sea urchins that prefer warmer water also moved in and devoured the common kelp. Huge expanses of kelp forests that were home to many marine creatures were wiped out.

The loss of kelp was only part of the problem. To escape warmer waters further north along the southern coast of Australia, many fish and other marine life moved south around Tasmania. But other fish that thrive in the previously cooler waters around the Tasmanian coast had no place to go. They cannot survive in the deep ocean waters closer to the South Pole because, similar to land animals, marine creatures belong in specific habitats. They are able to adapt to modest changes in their living conditions, but they cannot simply relocate to an entirely new environment. This is true of living things on land as well as in the sea.

Humans are the exception. While every other creature on the planet lives as part of nature, people erect buildings to create their own enclosed microhabitats, largely at the expense of the natural world.

Freshwater Fish

"Surely we all have a responsibility to care for our blue planet. The future of humanity and indeed, all life on earth, now depends on us."

– David Attenborough

Increasing water temperatures are not just harming the marine environment. Freshwater fish are feeling the pain as well. The internal temperature of freshwater fish is determined by the temperature of the surrounding water. For their bodies to operate most effectively, fish require water temperatures in a specific range, although a fish's tolerance for changes in water temperature varies from species to species. Water that is too warm can harm a fish's vital functions such as respiration, reproduction, growth, and buoyancy. In the past water temperature changes have been slow enough to allow fish to gradually evolve by adjusting their thermal limits.

Some fish can adapt to minor changes in their optimum temperature range. But as water temperatures continue to rise, many fish will have to find cooler water to survive. Some fish in rivers and streams could migrate north, and fish in lakes and smaller water bodies may acclimate to cooler, deeper water. But unlike the open ocean, streams and many lakes are confining and offer far fewer opportunities for fish to seek out their optimum habitat elsewhere when conditions deteriorate. For example, fish in a stream oriented primarily east-west are unable to move north to seek cooler water. As we've seen, saltwater fish, terrestrial mammals, and birds have their own challenges as temperatures rise, but in general, they can disperse more easily than freshwater fish to find suitable habitats.

While freshwater fish are harmed directly as warmer water raises their body temperature, additional disruptions to their habitat due to temperature increases are also taking a toll. Warmer waters alter the biological communities in lakes and streams including the type and abundance of prey for fish. Species more suited to higher temperatures move in. Native fish have to compete with these invasive species and can be crowded out. For example, native fish in the Great Lakes are already competing with fish and other aquatic life introduced by shipping and other human activities. Increasing water temperatures are compounding the problem by causing many species to expand their range northward. As native species in Lakes Ontario, Erie, and Michigan migrate north to Lakes Huron and Superior to find cooler water, they are likely to be replaced to some extent by new species that compete with the remaining native fish. At the same time, the fish that moved north compete with species already occupying their new environment.

Warmer freshwater, just like warmer saltwater, holds less oxygen essential to fish respiration. Fish become more prone to disease from the stress of warmer water, and algal blooms, some of which are toxic to fish, become more common. More frequent and prolonged droughts that are a consequence of greenhouse gas emissions reduce water flow and circulation and lead to stagnation and water quality degradation in fresh water bodies, creating conditions conducive to more algal blooms. When the algal blooms decay, they further reduce the amount of dissolved oxygen in the water.

Less mixing also means less dissolved oxygen in deeper waters. That can alter species distribution by replacing some fish with others that are more tolerant of low oxygen levels. Changes in circulation patterns in large water bodies also affect nutrient concentrations and upset the ecological interactions critical to fish.

Each of these consequences of increasing water temperature does not work independently. As in every ecosystem, natural processes in aquatic environments are interrelated. Changes in one parameter, such as water temperature, can set in motion fundamental, irreversible modifications to an entire aquatic community.

Into the Abyss

"Humankind has not woven the web of life. We are but one thread within it. Whatever we do to the web, we do to ourselves. All things are bound together.
All things connect."

– Chief Seattle

Planet Earth is alive. The world is made up of many complex communities of living plants and animals that sustain themselves by interacting with each other and their physical environment. It may appear as if each of these ecosystems is independent of the other, each doing its own thing. After all, rain forests, deserts, glaciers, and oceans don't have much in common. But ultimately, they're all connected. Greenhouse gas emissions are warming the entire planet, not discrete regions, continents, or oceans. Some ecosystems are more resilient than others, but ultimately every place on Earth will be harmed either directly by warming or indirectly by the connectedness of every part to the whole. There are no safe havens, no places to hide.

When a specific geographic area that characterizes a particular ecosystem, such as the Amazon rainforest, experiences an intrusion, it weakens its normal functioning. But natural occurrences, such as periodic fire, flooding, and drought, are part of the ecological process, and a large, resilient natural area can heal itself and recover from minor offenses. If the insult is overwhelming, however, such as relentless global warming, the stress to ecological processes can reach a point where the ecosystem can no longer compensate. It reaches a tipping point where the stress factors cannot be stopped and an irreversible breakdown commences. The area will have been injured so severely it cannot heal itself; it is, in effect, terminally ill.

Trees sustain the continuous cycle of water in a rainforest. Once rain saturates the soil, water is taken up by trees through their roots and then reenters the atmosphere through transpiration from leaves. The Amazon rainforest has been under attack by humans for decades. Vast expanses have been cleared for timber, farming, cattle ranching, and mining. Wildfires are also a factor in this accelerating destruction. Increasing air temperatures are making matters worse by drying soils and surface water and increasing transpiration and evaporation.

Deforestation in the Amazon may soon reach the point where there are no longer enough trees to maintain the cycle. The basic structure of the rainforest will collapse, and it will be transformed into an entirely different ecosystem, a savanna mostly devoid of trees. When that happens the indispensable function of storing carbon will be lost and the rich, diverse wildlife representative of the rainforest will disappear. The carbon released into the atmosphere from deforestation is adding to the carbon from greenhouse gas emissions, which is speeding up atmospheric warming. It is becoming a vicious, self-perpetuating cycle of accelerating global warming.

Greenland and the ice sheet that covers about 80 percent of it comprise a completely different ecosystem than the Amazon rainforest. But a similar critical point of irreversible transformation may be about to occur there because the excessive heat from greenhouse gas emissions could lead to unstoppable melting. The ice sheet is huge, covering over 600,000 square miles. Most of it is more than a mile thick and it contains enough water that, if it all melted, it would raise the planet's sea level by more than 20 feet.

Rising air temperature is not the only cause of the accelerating melting of the ice sheet. During summer algae grow on the surface of the glaciers. The algae are darker than the ice, often green or brown, so it absorbs more heat than the bright surface of the ice and increases the melting. Most of Greenland has been covered with ice for millions of years, and the extent of the ice fluctuates in response to the rhythms of our living planet. But just as in the Amazon rainforest, the physical and biological processes in Greenland that typically keep this complex natural system in check are being overwhelmed by an aggressive intruder, unrestrained global warming.

Only several decades ago the assumption was that global warming was happening gradually, a steady increase in atmospheric temperatures, something humans could predict, plan for, stop, and mitigate. Now it's understood that things are much messier. In addition to more intense hurricanes, floods, droughts, and other extreme weather events, disruptions to the world's ecosystems are happening in unanticipated bursts. It appears that this erratic behavior may be occurring just as an ecosystem, such as the Amazon rainforest or Greenland, is about to reach its tipping point.

Ocean currents are the threads that tie together ecosystems around the planet. If a thread loosens or breaks, it causes disconnects and disruption of basic natural functions. Warming of the air/water interface, along with other factors such as excessive freshwater from melting ice sheets pouring into the oceans, are altering ocean circulation, slowing and shifting the paths of major currents that spread heat around the globe. Complicated relationships between these physical processes and biological systems are having a cumulative effect that produces unpredictable transitions in ecosystems. The engine driving all this disruption is increasing greenhouse gas emissions.

"When one tugs at a single thing in nature, he finds it attached to the rest of the world."

– John Muir

Yes, You Can

"Climate change is real. It is happening right now, it is the most urgent threat facing our entire species and we need to work collectively together and stop procrastinating."

– Leonardo DiCaprio

One sunny day I was minding my own business when out of the corner of my eye I saw a hawk circling above. Henry and the other crows weren't around to sound the alarm, so I raised a racket. I have this whole repertoire that I sing out to alert my neighbors when trouble is around. They know right away what I'm saying. It really annoys the hawks and other predators, and I understand that they need to eat just like the rest of us. But I draw the line when they look at my friends as lunch.

The point is, I'm just a little bird, but I did something. I made a difference. You can, too. You don't need to be a superstar to help save the planet. What's important is that each of you does something. You can make a difference. If you'll do that, I promise I'll eat out of your hand. Um, sunflower seeds, please.

Epilogue

Human Nature

"How we humans came to be the way we are is far less important than how we should act now to get out of the mess we have made for ourselves."

– Jane Goodall

Humans have been a blip in time. If they continue their destructive ways they may be gone soon. They brought this on themselves.

Human nature includes a bundle of inherited traits. People come into this world with the ability to behave and think in various ways depending on which traits are suppressed or expressed. There are over seven billion humans on Earth, and they certainly don't all behave and think the same way. Individuals are different because from the day they are born they are shaped by their environment and upbringing, as well as by choices they make along the way.

Humans are causing the warming that is devastating the planet, and only they can fix it. But to do so requires that many of those billions of people take positive, decisive action. Sadly, for too many people, one or more inherent traits have dominated during their formative years that make these people ineffective and, in some cases, destructive.

Too many people are self-centered. They have an "it's-all-about-me" attitude. Beyond their immediate family, they don't seem to care. They fail to see that by helping others they would help themselves. They don't understand that they need to personally contribute and work with others to make life better for everyone. They may recognize the seriousness of global warming but feel no obligation or incentive to do something about it.

Many people are conceited. They think they have all the answers. They're obsessed with their perceived brilliance and expect that the next generation, or the one after that, will come up with some clever technological gadget that will stop the warming, and the planet will magically heal itself. Well, look where that's gotten us. Humans have made a mess of things with their total disregard for the home all of us share.

People can be cruel. They are not nice to each other, let alone to the world around them. They lack compassion. When you're compassionate, you do good things. But many humans spend a lot of time hating and hurting each other. If they lack compassion for their fellow human beings and for life on the planet, they're not about to take action to save us all from catastrophe.

Finally, many people are superficial. They mostly care about stuff, not substance, not what truly matters. But having more things won't make life better if people make the planet, everyone's home, unlivable. Humans are always running about. They need to pause, take a deep breath, and smell the flowers. If people took the time to reflect on the significance and beauty of nature and the damage they are causing, they would be more likely to do something about it.

Over millions of years, the world has changed and many creatures have come and gone. But wild creatures are not self-destructive by nature. Humans, it seems, are different. How ironic that they call themselves civilized when in fact many of them – especially the ones in charge – are complacent, crude, and callous. Too many people settle for apathy rather than empathy, hatred rather than love. But not all humans suffer this affliction. We must hope that those humans who listen to the better angels of their nature will prevail.

Into the Sunshine

"We are the first generation to feel the sting of climate change, and we are the last generation that can do something about it."

– Jay Inslee

The next generation
Capt. John Severns, U.S. Air Force

The word among our other feathered friends is that chickadees are audacious. We're too trusting of humans. People love to watch us eat sunflower seeds from their outstretched hands. These are people like you who respect the natural world, and you are the ones who must save it.

In closing, I want to make a personal appeal to my human neighbors. To be blunt, you got us into this mess and you are the only creatures on the planet who can get us out of it. Our young and yours deserve a future. This problem will not be fixed by some brilliant scientists and engineers providing a technological solution to a handful of altruistic world leaders who then agree to make it all better. If that were the case, it would have happened already. It hasn't. In fact, in spite of all the dire warnings, greenhouse gas emissions continue to rise globally.

Decision makers will only take decisive action when millions of people just like you demand it. Each of you needs to act, you need to do it right now, and no matter what the obstacles, you must never give up. If things look hopeless, turn to nature for solace, and remember:

The Peace of Wild Things

When despair for the world grows in me
and I wake in the night at the least sound
in fear of what my life and my children's lives may be,
I go and lie down where the wood drake
rests in his beauty on the water,
and the great heron feeds.
I come into the peace of wild things
who do not tax their lives with forethought
of grief. I come into the presence of still water.
And I feel above me the day-blind stars
waiting with their light. For a time
I rest in the grace of the world, and am free.

– Wendell Berry

That's my story. I've gotta go. I promised Aldo and Henry I'd meet them at the bird feeder. There's a light breeze and the sun is shining. It's a beautiful day!

With love and hope,

Greta

www.ingramcontent.com/pod-product-compliance
Lightning Source LLC
Chambersburg PA
CBHW040518220526
45473CB00012B/2908